RUBBER BAND
ENGINEER

ALL-BALLISTIC Pocket Edition

RUBBER BAND
ENGINEER

ALL-BALLISTIC Pocket Edition

FROM A SLINGSHOT RIFLE TO A MOUSETRAP
CATAPULT, BUILD 10 GUERILLA GADGETS
FROM HOUSEHOLD HARDWARE

ROCKPORT

LANCE AKIYAMA

Brimming with creative inspiration, how-to projects, and useful information to enrich your everyday life, Quarto Knows is a favorite destination for those pursuing their interests and passions. Visit our site and dig deeper with our books into your area of interest: Quarto Creates, Quarto Cooks, Quarto Homes, Quarto Lives, Quarto Drives, Quarto Explores, Quarto Gifts, or Quarto Kids.

Rockport Publishers titles are also available at discount for retail, wholesale, promotional, and bulk purchase. For details, contact the Special Sales Manager by email at specialsales@quarto.com or by mail at The Quarto Group, Attn: Special Sales Manager, 100 Cummings Center, Suite 265-D, Beverly, MA 01915, USA.

10 9 8 7 6 5 4 3 2 1

ISBN: 978-1-63159-738-1

Digital edition published in 2019
eISBN: 978-1-63159-739-8

Library of Congress Cataloging-in-Publication Data originally found under *Rubber Band Engineer*.

Cover Design: Laia Albaladejo
Design and Page Layout: revised by Laia Albaladejo based on Timothy Samara's original design
Photography: Lance Akiyama
Illustration: Timothy Samara

Printed in China

Thank you to my parents, Nancy and James, for allowing me to take apart old electronics and for letting me turn my childhood bedroom into a wonderful, creative mess. And, thank you to the inventors of rubber bands, cardboard, and tape, without whom I might not have a career.

LANCE

CONTENTS

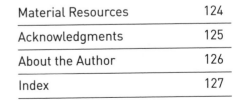

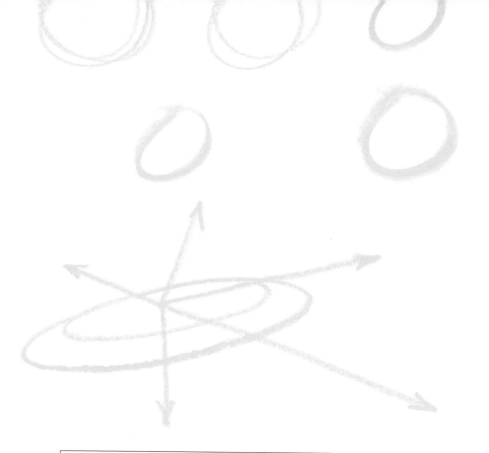

INTRODUCTION

$$v^2 - v_0^2 = 2a\Delta x$$

$$v \; \frac{v + v_0}{2}$$

Do you have a rubber band handy? Good. As soon as you pick it up, you'll probably want to start playing with it. You have a choice: You could just pull it back on your fingertips and zing it across the room, but that's not why you have this book. You're a maker and a junk-drawer engineer, and you know that rubber band is destined for greater things, like a PVC Slingshot Rifle or a Pyramid Catapult.

This new pocket-sized, all-ballistic edition of *Rubber Band Engineer* is put together just for makers like you. Toss it into your backpack, your bike pack, your shoulder bag, or your glove compartment. You never need to be without it when the irresistible urge for a little guerilla engineering grabs you.

Of course, you know that there's more to life than just rubber band-powered shooters. We've got you covered. These ten projects range from simple darts to complex contraptions such as a Crossbow or a Floating Arm Trebuchet, in which you'll harness the potential energy of paint stirrers and the awesome power of gravity itself.

Grab a glue gun, clear a workspace, and prepare for the mild amount of mayhem that is sure to ensue. It's time to start making!

Lance Akiyama

HANDHELD SHOOTERS

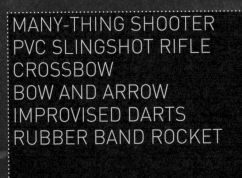

MANY-THING SHOOTER
PVC SLINGSHOT RIFLE
CROSSBOW
BOW AND ARROW
IMPROVISED DARTS
RUBBER BAND ROCKET

MANY-THING SHOOTER

$$\Delta x = \bar{v}t$$

$$\Delta x = \frac{1}{2}at + v_0t$$

$$v = at + vo$$

The Many-Thing Shooter earns its name from its versatility: It can be built from many things. It can also launch many things, including candy, beads, and pieces of cork, to name a few. Every material in the shooter can be replaced with another household item, making this one of the most imaginative, adaptable, and quick-to-build projects in this book. Experiment with different projectiles and rubber-band combinations until your Many-Thing Shooter becomes your go-to DIY sidearm.

 CHOOSE YOUR RUBBER BAND WISELY

The rubber band is a critical component of the shooter. Choose a band that has a maximum stretched length about equal to the length of the stick. Avoid using very thin bands that break easily or very thick bands that are difficult to stretch. Try doubling your rubber bands for additional power!

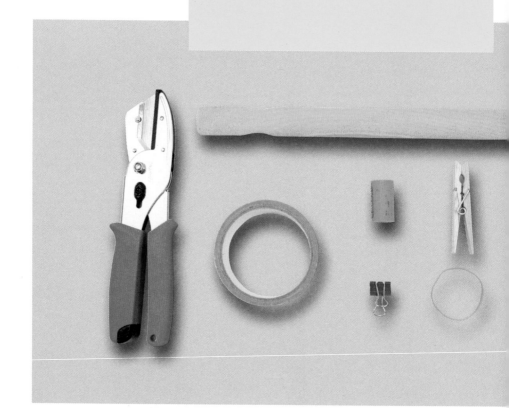

TOOLS + MATERIALS
DUCT TAPE
SCISSORS
PAINT STIRRER
SPRING-TYPE CLOTHESPIN
RUBBER BAND
SMALL BINDER CLIP
BOTTLE CORK (OR PROJECTILE OF YOUR CHOICE)
CUTTING TOOL (OPTIONAL)

MATERIAL SUBSTITUTIONS	
PAINT STIRRER	Ruler, wood shim, craft sticks; or another flat, rigid strip of wood, plastic, or cardboard
CLOTHESPIN	Binder clip, chip clip, or anything that can clamp onto a rubber band and act as a trigger
BINDER CLIP	Masking tape or duct tape
DUCT TAPE	Hot glue, masking tape, or any glue or tape that can secure the trigger in place
PROJECTILE	Anything that is slightly wider than the trigger. Reasonably dense and aerodynamic objects, such as wooden beads, pebbles, or candy work best.

01 > Make the trigger. Cut a 6" (15 cm) piece of duct tape and split it lengthwise. Use the two pieces of tape to attach the clothespin to one end of the paint stirrer, making sure the pinching tip of the clothespin faces the middle of the stick.

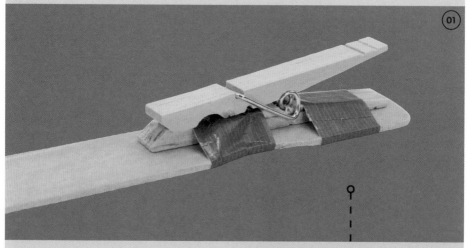

02 > Use the binder clip to clamp the rubber band to the other end of the stick. Fold the binder clip handles down. Using a binder clip allows you to quickly swap out rubber bands with different shooting power or replace broken ones.

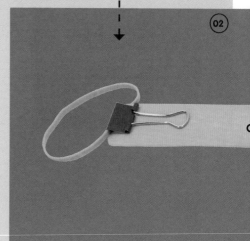

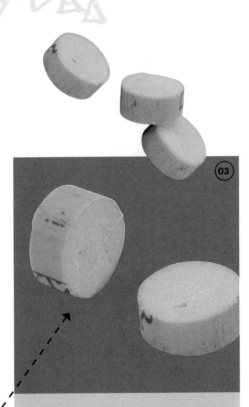

03 > Choose your projectile. For indoor shooting, try disks made by cutting a synthetic cork into quarter slices approximately ¼" to ½" (6 mm to 1.3 cm) thick. These corks are dense enough to shoot a good distance and maintain a fairly accurate trajectory but not so dense that you'll break a window. Ideally, the projectile should be slightly wider than the trigger so the rubber band can hold it in place.

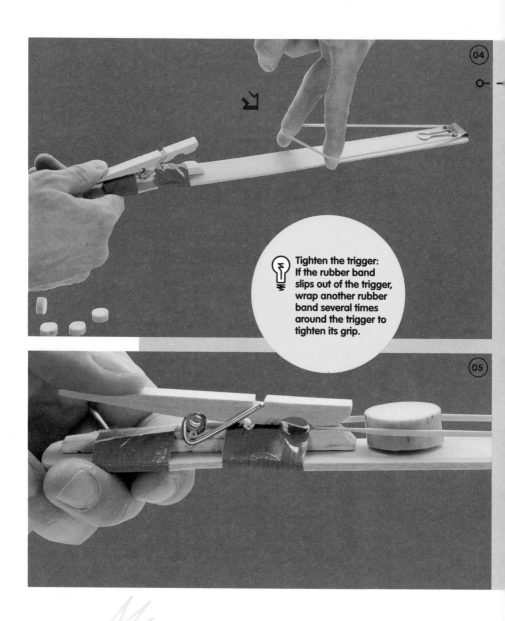

Tighten the trigger: If the rubber band slips out of the trigger, wrap another rubber band several times around the trigger to tighten its grip.

04 > You're ready to shoot! Squeeze the trigger open with one hand. Use two fingers from your other hand to stretch the rubber band into the open trigger. The rubber band should remain stretched taut when you clamp the trigger onto it.

05 > Wedge the projectile between the two sides of the rubber band directly in front of the trigger. This will prevent the projectile from falling out as you prepare your shot and ensures that the projectile receives the full force of the rubber band's elastic energy.

06 > Scan your surroundings for something fun to shoot at. Things that fall over, make noise, or shatter (so long as they're not of value to anyone!) are great choices. Avoid shooting point blank at hard surfaces—your projectile might ricochet back at you.

DESIGN VARIABLE

The shooter can be modified to fire unsharpened pencils. To prevent the pencil from veering to the side when it's fired, add a paper clip guide to the shooter. Open the paper clip into a right angle with one curved end wide enough for the pencil to pass through. Attach the paper clip to the shooter with tape.

NOW GET TINKERING

The Many-Thing Shooter can be attached to almost any flat and rigid surface. Try turning boring objects, like a binder or instruction manual, into an improvised firearm.

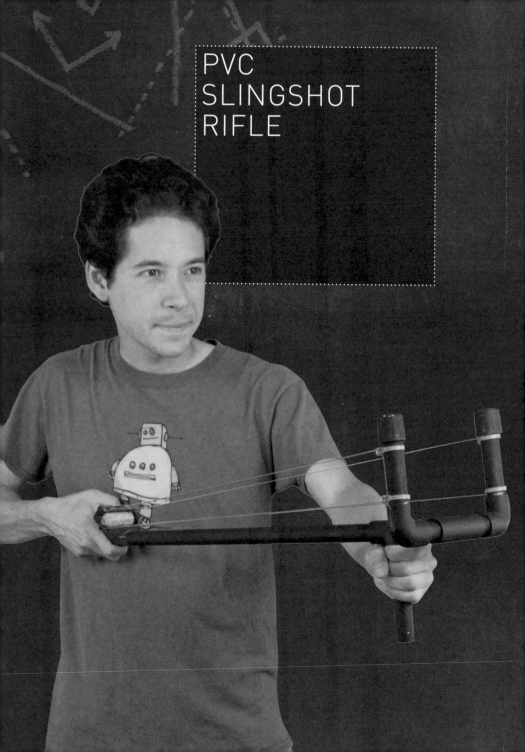

PVC
SLINGSHOT
RIFLE

(i) This high-powered slingshot
is incredibly accurate and reliable, yet it's
very simple to build and to modify.
Unlike typical slingshots, this one comes
with a satisfying triggered release, allowing
you to effortlessly line up your shot before
letting loose. This is what rubber bands
were made for!

PVC PIPE CUTTER OR HACKSAW	**PVC PRIMER**
42" (1 M 6.5 CM) OF ½" (1.3 CM) SCHEDULE 40 PVC PIPE	**PVC CEMENT**
	BLACK SPRAY PAINT (OPTIONAL)
TWO ½" (1.3 CM) PVC ELBOW CONNECTORS	**4 CABLE TIES**
	TWO 7" (18 CM) RUBBER BANDS
TWO ½" (1.3 CM) PVC TEE CONNECTORS	**CARDBOARD TUBE**
	DUCT TAPE
TWO ½" (1.3 CM) PVC END CAPS (OPTIONAL)	**LARGE BINDER CLIP**
	2 LARGE CRAFT STICKS
	SAFETY GLASSES

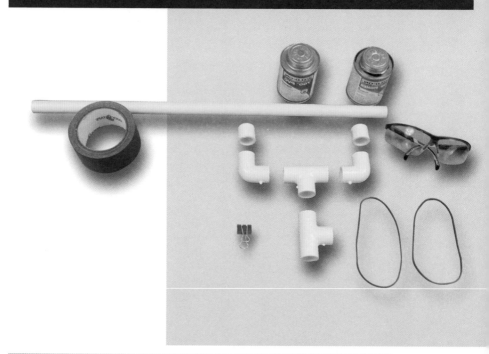

$$v^2 - v_0^2 = 2a\Delta$$

$$v \; \frac{v + v_0}{2}$$

$$g = 10\,\frac{m}{s^2}$$

MATERIAL SUBSTITUTIONS	
PVC PIPE CUTTER	**Hacksaw or another type of saw**
7" (18 cm) RUBBER BANDS	**A chain of shorter rubber bands**
CARDBOARD TUBE	**Any scrap of cardboard**
LARGE CRAFT STICKS	**Any rigid piece of wood or plastic of a similar size**

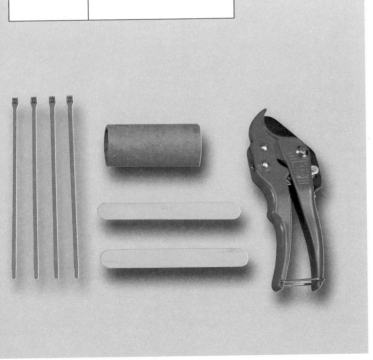

01 > With the pipe cutter, cut three 2" (5 cm) lengths and three 4" (10 cm) lengths from the 42" (1 m 6.5 cm) PVC pipe. Set aside one 4" (10 cm) piece for the slingshot's grip and the remaining 24" (61 cm) length of pipe for the handle.

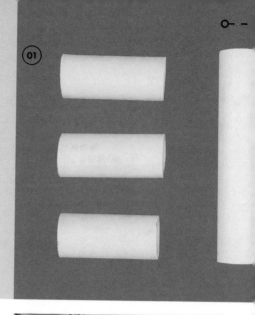

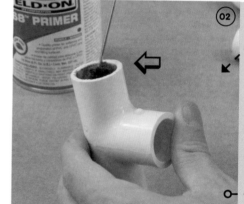

02 > Choose a well-ventilated area in which to work and protect your work surface from the primer and cement before starting. Apply primer and then cement to both ends of each piece of pipe and inside the openings of the connectors. Assemble the other short pieces of pipe, the connectors, and the end caps by simultaneously pushing and twisting the components together.

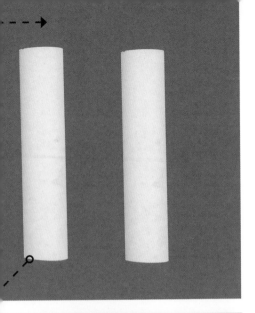

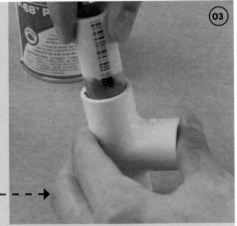

03 > Insert the 4" (10 cm) pieces of pipe into one end of each elbow connector. Use a 2" (5 cm) piece of pipe to connect each of the elbows to a tee connector. Use the remaining 2" (5 cm) piece of pipe to join the two tee connectors, turning them perpendicularly to each other. **Optional:** Top off the two 4" (10 cm) pieces of pipe with the end caps.

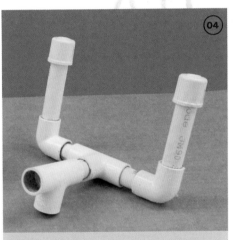

04 > Your slingshot will look like this. Allow the cement to set according to the manufacturer's directions.

05 > Affix the remaining 4" (10 cm) piece of pipe to the base of the slingshot as a grip. Attach the 24" (61 cm) piece of pipe to the tee connector for the handle. These two pieces of pipe don't require cementing.

DESIGN VARIABLE

Cemented PVC pipe has an unpolished appearance. It's apt to be covered in logos, numbers, bar codes, and stains from runny PVC primer. Spray paint your slingshot to cover up the blemishes; it won't impact performance, but it will look a lot more devastating.

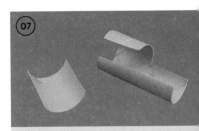

07

07 > Cut a 2" × 3" (5 × 7.5 cm) rectangle from the cardboard tube. This will form the slingshot's "sling." If you're using plain cardboard, roll it so it curves. You could also fashion a sling from just duct tape, but the tube offers a helpful premade curved shape.

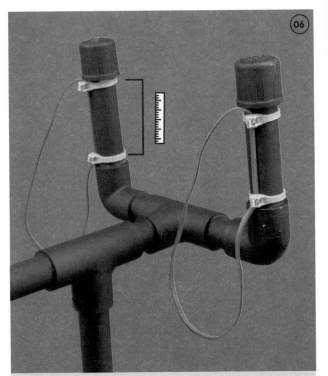

$v = at + vo$

06 > Strap both ends of the rubber bands to the slingshot wye with cable ties. The cable ties should be spaced at least 3" (7.5 cm) apart. This configuration prevents the rubber bands from twisting upon release.

08 > Position the rubber bands around the curve of the sling. Secure the sling to the rubber bands with duct tape, and then test it by pulling it back. There should be even tension in each of the four rubber-band strands.

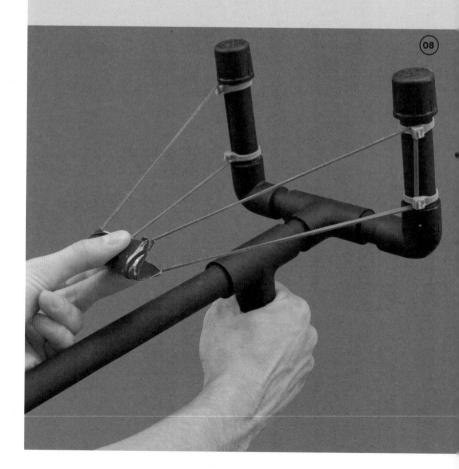

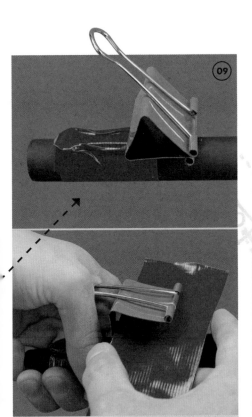

09 > Make the trigger. Attach the large binder clip to the end of the slingshot handle with duct tape. Press the binder clip open. Further secure the trigger by wrapping tape around the slingshot handle and the inside of the binder clip.

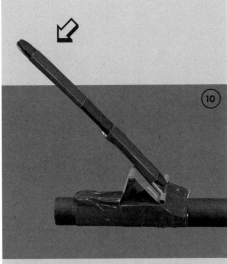

10 > Stack the two craft sticks and bind them together with duct tape. Attach the sticks to the trigger's upper binder-clip handle with tape. This will give you more leverage, allowing you to release the trigger easily.

11 > Test fire! Load your projectile into the sling. Small, round, evenly weighted objects like corks and hard candy work best.

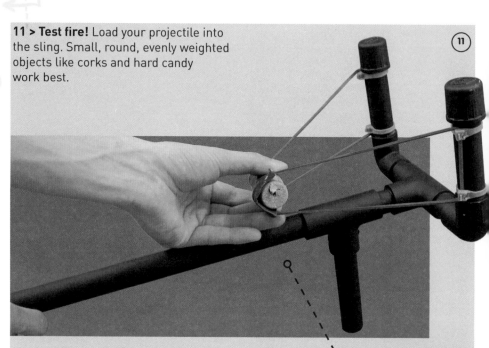

⑪

12 > Pull back on the loaded sling. Open the trigger and insert the sling. The slingshot is loaded.

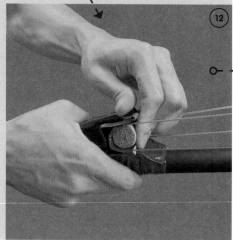

⑫

This design is very reliable; however, you should always wear protective eyewear when firing a homemade slingshot. There is a chance that a sling can backfire or that the projectile will ricochet. Never aim by holding the back of the slingshot up to your eye; fire from the hip.

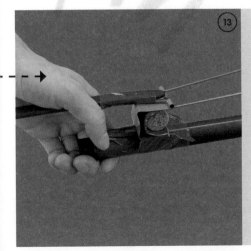

(13) **13 > Fire away!** Release the sling by grasping the trigger handle.

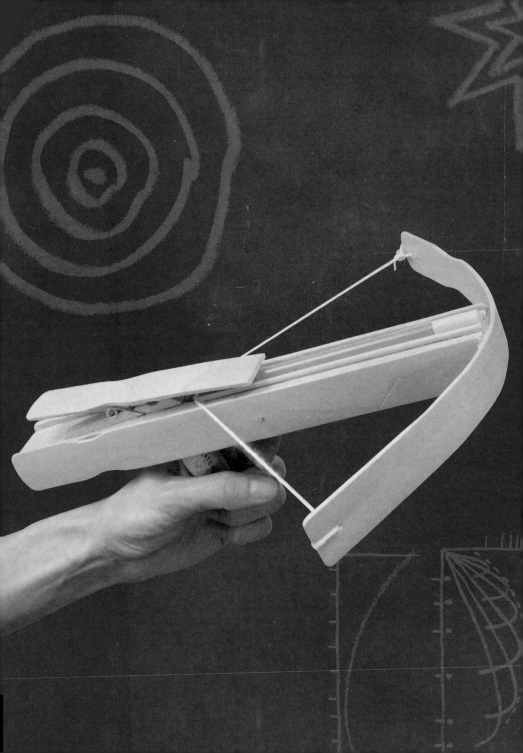

CROSSBOW

(i) This project requires some determination to complete, but the payoff is worth it. The crossbow makes a satisfying *snap* when the trigger is pulled, and it can launch bolts more than 100 feet (30 m)! You can power this crossbow with a piece of string or with a rubber band.

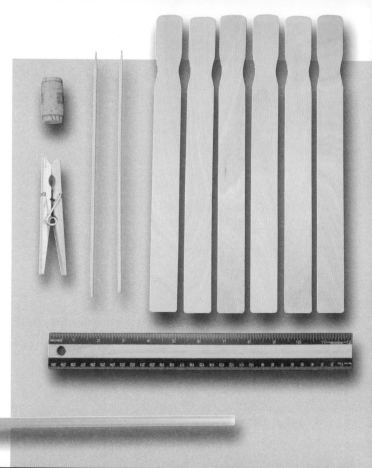

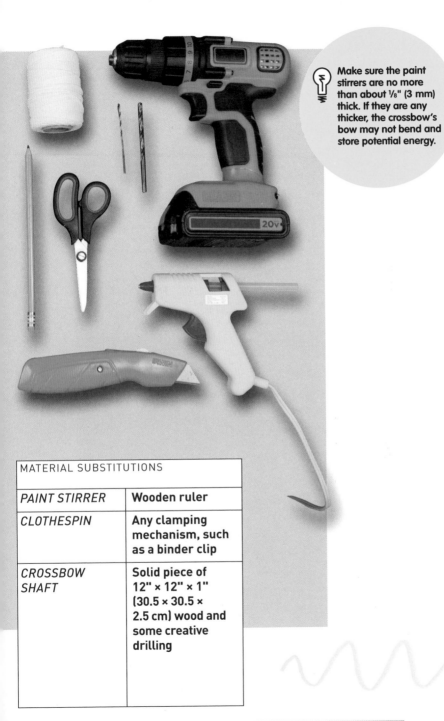

Make sure the paint stirrers are no more than about ⅛" (3 mm) thick. If they are any thicker, the crossbow's bow may not bend and store potential energy.

MATERIAL SUBSTITUTIONS	
PAINT STIRRER	**Wooden ruler**
CLOTHESPIN	**Any clamping mechanism, such as a binder clip**
CROSSBOW SHAFT	**Solid piece of 12" × 12" × 1" (30.5 × 30.5 × 2.5 cm) wood and some creative drilling**

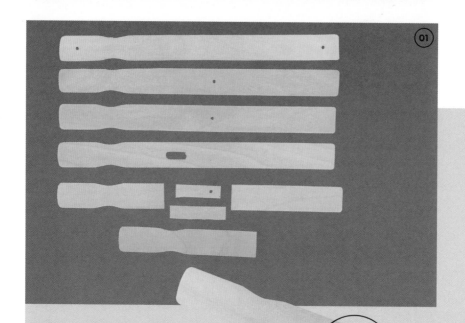

01 > Begin by preparing the paint stirrers as shown here. For the first, drill a ⅛" (3 mm) hole near each end of the stick. For the second and third, drill a ⅛" (3 mm) hole that is 5¼" (13.5 cm) from one end of each stick.

02 > For the fourth paint stirrer, using the ⅜" (1 cm) bit, drill a hole 4¾" (12 cm) from one end. Lift the drill, move it toward the center of the stick, and drill a second hole ½" (1.3 cm) from the first.

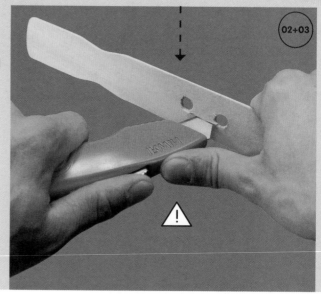

02+03

04 > For the fifth paint stirrer, use the utility knife to score and break the stirrer into two 4¾" (12 cm) pieces and one 2" (5 cm) piece. Scoring the wood with the blade as described in step 3, split the 2" (5 cm) piece in half lengthwise. Trim ¼" (6 mm) off one of the 2" (5 cm) pieces and drill a ⅛" (3 mm) hole near the end of it.

03 > Using the utility knife, carve out the section between the two holes created in step 2. Carefully cut along the grain of the wood by slowly pushing the knife away from your body and hands.

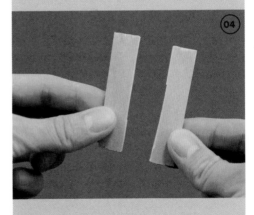

05 > Cut the sixth paint stirrer in half, crosswise. You'll only need one of the halves.

06 > Begin assembling the shaft of the crossbow like a box. Line up the paint stirrer from step 3 with one of the stirrers from step 1 with a hole 5¼" (13.5 cm) from the end. Take careful note of the orientation of the holes in the photo.

07 > Use hot glue to attach the edges of the paint stirrers at right angles. If the glue dries too quickly when you run a thin line of it along the edge, use several beads of glue instead.

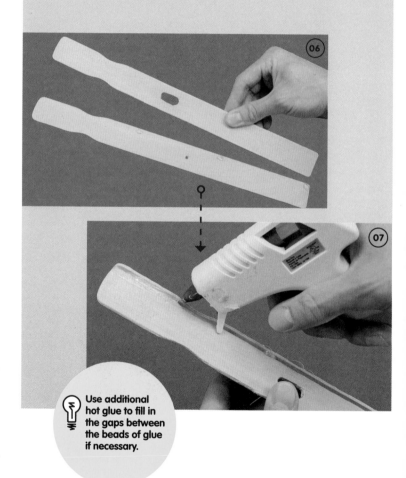

Use additional hot glue to fill in the gaps between the beads of glue if necessary.

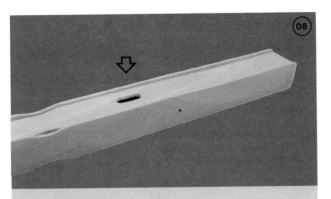

08 > Glue the second stirrer with a hole 5¼" (13.5 cm) from the end opposite the first. Make sure that the two ⅛" (3 cm) holes line up.

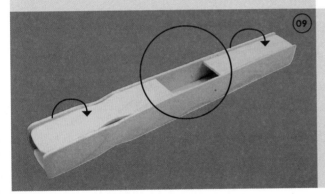

09 > Turn the box over and glue the two 4¾" (12 cm) stirrer pieces to the underside of the shaft. The 2" (5 cm) gap in the center will be where the trigger goes.

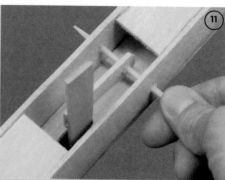

10 > **Make a trigger** with the two small pieces of wood from step 4. Position the piece with the ⅛" (3 mm) hole at a right angle to the other, ½" (1.3 cm) down from the top. Hot glue the two pieces together.

11 > Set the trigger into place. Choose the thickest skewer and thread it through the holes in the crossbow shaft and trigger. This design relies on the friction between the skewer and the ⅛" (3 mm) holes to hold it in place. The ½" (1.3 cm) of the trigger's crosspiece should poke through the cut-out slot.

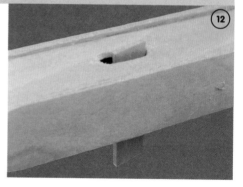

12 > This is how the trigger should look, and it should swing up and down. If it doesn't quite fit, then you may need to drill or carve out a larger hole or drill new holes for the trigger hinge. When you have it working, trim off the extra lengths of skewer and save the scraps.

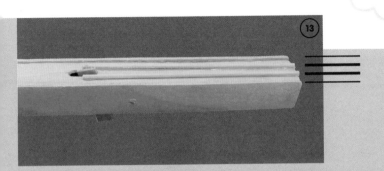

13 > Create the guides for the bolt.

Cut the skewer scrap and the second skewer to 6¾" (17 cm) lengths. Glue the two skewer pieces parallel to each other with about ⅜" (1 cm) between them. The ends of the skewers should line up with the front of the crossbow shaft and the middle of the trigger head. Make sure the ends of the skewer nearest the trigger have a clean and flat cut, or the string might not latch on correctly.

14 > Center and glue the paint stirrer with the ⅛" (3 mm) holes drilled at each end to the front end of the crossbow shaft.

15 > Thread the string through the ⅛"
(3 mm) holes and knot the ends. To make
the knot tying easier, an 18" (45.5 cm)
length of string is recommended,
but ultimately the string should be 12"
(30.5 cm) long from one hole to the other.
The bow should start to bend as you pull
the string back a few inches.

Here is a crossbow outfitted
with a rubber band, which
may be the way you'd like
to go. Getting just the right
string length is a trial-and-
error process. Rubber bands
are easier to load and easier
to calibrate to achieve the
right amount of energy, but
they may not provide as
much force as a taut string.

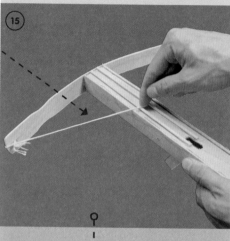

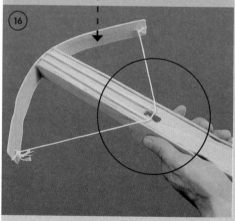

16 > Give the string a test by pulling it
back and slipping it over the ends of the
skewers. (If this is difficult to do with
your fingers, use the pencil's eraser to
push the string into place.)

You may need to calibrate the tension
of the string to achieve the most force.
The string should be a little loose when
under no tension, but very taut when
loaded. If you are unable to load the
string at all, then it needs to be a little
longer. Trim the ends of the string with
scissors when complete.

Note: If the string is slipping off of
the skewers, use a utility knife to
cut each skewer end, either flat or
slightly indented.

17 > Glue the clothespin directly behind the cut-out slot.

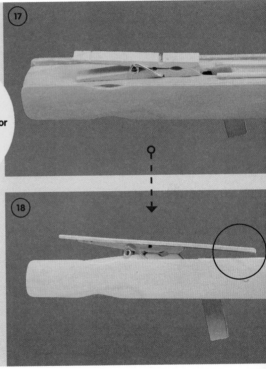

> **You can use the clothespin as a simplified trigger for the rubber band–based design.**

18 > Glue the half piece of paint stirrer from step 5 onto the top of the clothespin. The lower end of the stirrer should be between ¼" and ⅜" (6 mm and 1 cm) from the top of the shaft. This will be used to hold bolts in place.

19 > Test the trigger.

Hot glue the cork approximately 1½" (4 cm) behind the trigger. Pulling on the trigger should push the string off the ends of the skewers. If the string gets caught on one skewer, double-check to make sure that the trigger is centered exactly between the skewers. Also check to make sure that the ends of the skewers are about the same diameter and flat.

Create Crossbow Bolts

Now it's time to create the crossbow bolts. There are many options for creating ammunition: pens, pencils, dowels, and even hard candy will work. These bolts, made from drinking straws, are designed for distance.

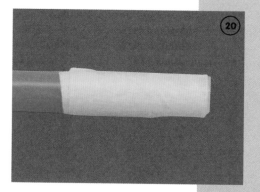

20 > Cut a thick drinking straw to 6" (15 cm) in length. Insert something dense in the tip, like a 2" (5 cm) piece of a hot glue stick. This piece of glue is called a leading weight. (See the notes on leading weights on page 52.)

Tape the glue stick in place and wrap tape around the other end of the straw.

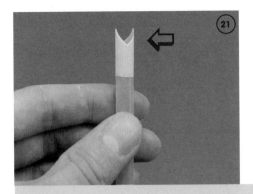

21 > Cut a nock into the back of the straw by pinching the tip until it's flat, then cutting off the corners. A nock will ensure that the bolt engages the string and fires consistently.

22 > Get ready to fire! Load the string behind the skewers on the crossbow. Place your bolt directly in front of the trigger hole, but don't cover it. Close the bolt holder on top of the bolt to keep it in place while you prepare your shot. Fire when ready.

DESIGN VARIABLES

This is just one example of a crossbow, but you can scale the proportions of the crossbow frame to be larger or smaller.

Do a quick internet search for "crossbow trigger diagram" to find more complex ways of releasing the string.

Add a thumbtack to the tip of your bolt and set up some cardboard targets.

Create a compartment for holding your ammunition.

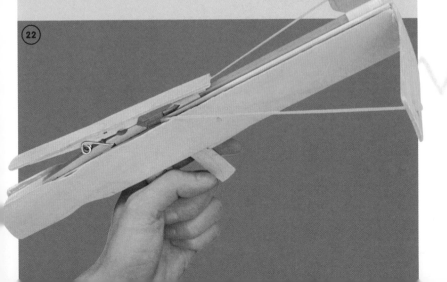

BOW AND
ARROW

ⓘ Branches, sticks, string, rubber
bands, pens, rulers, and bamboo skewers
can all be used to craft a bow and arrow.
This design is one way to do it, and it
illustrates the important principles. Like all
the projects in this book, it's built to be as
fun and effective as it looks.

2 YARDSTICKS (OR METER STICKS)
DUCT TAPE
UTILITY KNIFE
PLIERS (OPTIONAL)
STRING
DOWEL, 24" (61 CM) LONG
PENCIL SHARPENER (OPTIONAL)
SPOOL OF BENDABLE, NON-ALUMINUM WIRE, AT LEAST 18 GAUGE

MATERIAL SUBSTITUTIONS

YARDSTICKS	**Tape several paint stirrers together in an overlapping pattern; any strong yet flexible material that can be cut to the dimensions of a yardstick**
STRING	**A long chain of linked rubber bands**
DOWEL	**Bamboo gardening stakes**

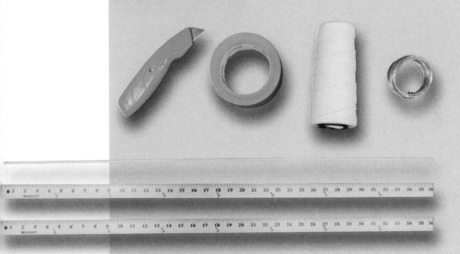

01 > Stack one yardstick (meter ruler) on top of the other and tightly wrap the ends together with duct tape. If your yardsticks have a hole at one end, make sure that both holes line up. If there is not a hole, carefully drill one with a ¼" (6 mm) drill bit, or create another notch as shown in step 2 and simply loop the string around it as in step 3.

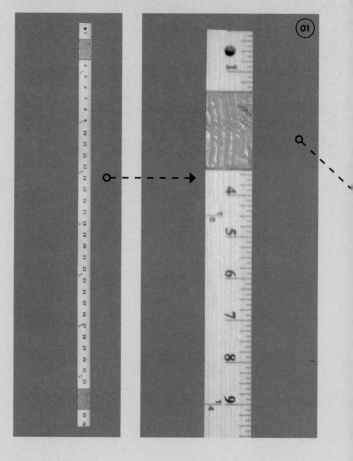

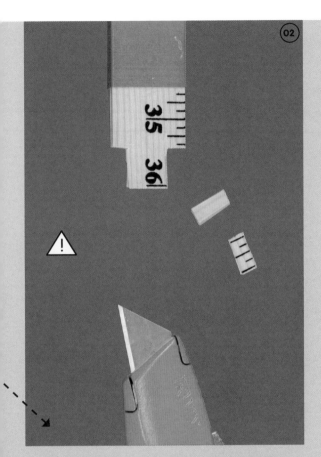

02 > Cut a ¼" × ½" (6 mm × 1.3 cm) piece off each corner of the yardstick. Use the utility knife to score the yardstick several times and then break the piece off. Because the grain of the wood runs lengthwise along the yardstick, you should have a clean break. Use pliers to snap off the corners if you have trouble breaking the wood.

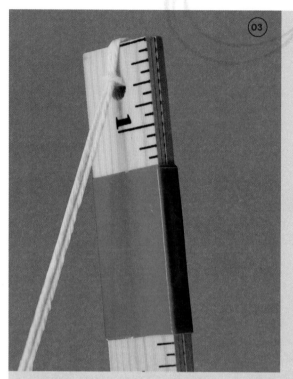

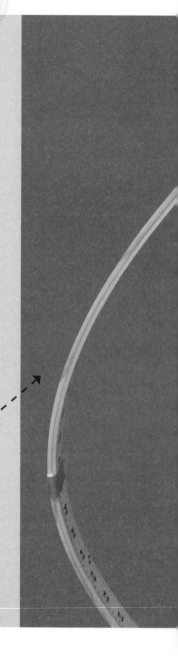

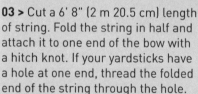

03 > Cut a 6' 8" (2 m 20.5 cm) length of string. Fold the string in half and attach it to one end of the bow with a hitch knot. If your yardsticks have a hole at one end, thread the folded end of the string through the hole.

04 > String the bow by bending it significantly while holding the loose ends of the string in your hand. Wrap the string around the second end of the bow and knot it.

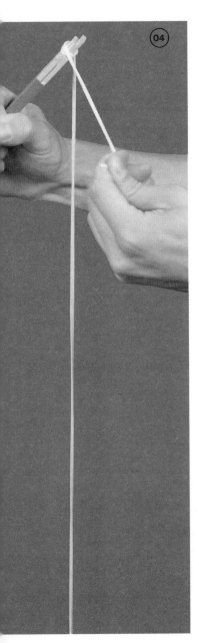

05 > Cut a 1" (2.5 cm) piece from the dowel. Tape the 1" (2.5 cm) piece of dowel so that it lines up with the very center of the bow. (If you are using yardsticks, this is at the 18" mark [50 cm on a meter ruler]). This will serve as the arrow rest.

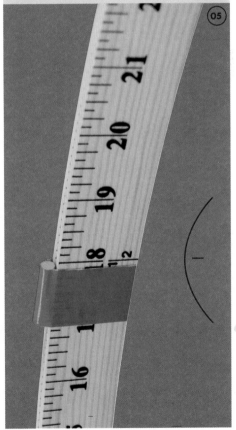

06 > Now turn the long piece of dowel into an arrow. Use the utility knife to cut a V-shaped nock at one end of the dowel. The nock will prevent the arrow from slipping off the bow string when you shoot.

THE IMPORTANCE OF LEADING WEIGHTS

Imagine trying to throw a long strip of paper. It won't go very far and it definitely won't go straight. Now imagine attaching a rock to one end of the paper and throwing it again. The momentum of the rock will carry the paper through the air and the paper will trail behind it in a straight line. This same idea applies to long projectiles—such as arrows. They need weight at the tip in order to work.

07 > Use the utility knife to whittle the other end of the dowel to a point. Alternatively, use a pencil sharpener for a more precise point.

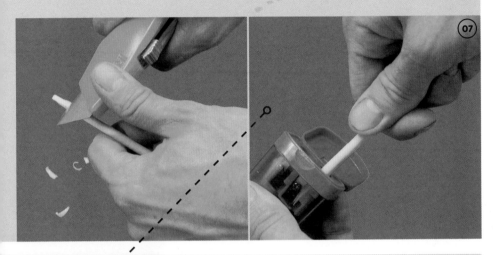

08 > Wrap fine wire around the dowel near the arrow's tip to create a 1" (2.5 cm)-wide band. Adding this leading weight will allow the arrow to fly straight.

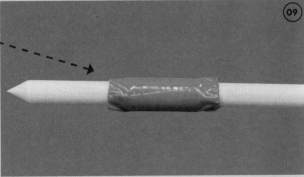

09 > Wrap the wire with duct tape to secure it. You're ready to fire!

10 > Nock the arrow onto the string and rest the shaft on the arrow rest. Contrary to the techniques of real archery, you will pinch the nocked dowel and draw it back until only the very tip of the arrow is in front of the bow. Aim safely, and let go! Sheets of cardboard make great targets.

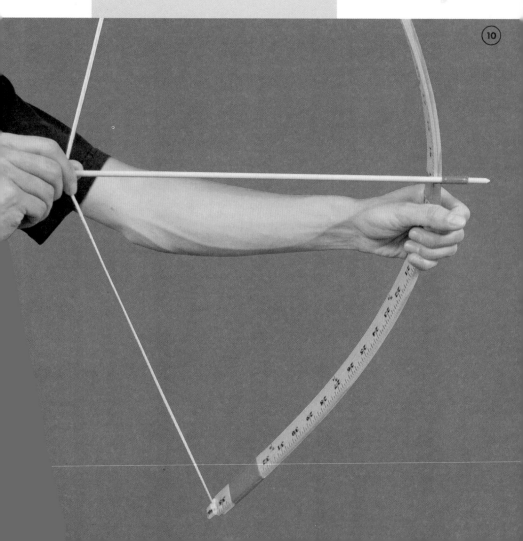

WAIT, WHERE'S THE FLETCHING?

Real arrows use fletching (fins or feathers) in addition to a leading weight to stabilize the arrow's flight. Arrows made from household materials sometimes don't. Here's why:

Real arrows are flexible. If you watch an arrow being shot in slow motion, it will bend significantly under the force of the released bowstring. This bending allows the arrow to effectively curve around the bow before straightening out in midflight. This is important!

If the arrow did not bend, the fletching would collide with the bow and be ruined. That is exactly what will happen if you add fletching to this inflexible arrow. If you make an exceptionally strong bow, or you use a more flexible material for the arrow, your arrows will benefit from fletching. Want to know more? Do an internet search for "the archer's paradox."

Darts may be the simplest project in this book, but don't be fooled—creating a super-effective dart is more nuanced than you might think. This project isn't a step-by-step plan, but rather a guide to illustrate what factors contribute to—and detract from—a great improvised dart, whether made from a pen, a pencil, or a drinking straw.

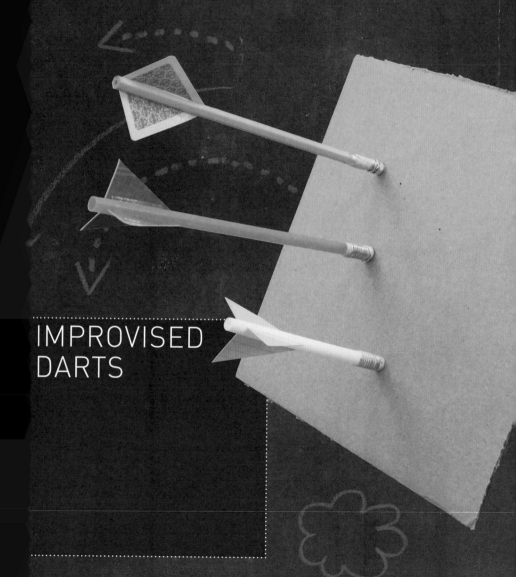

IMPROVISED DARTS

Pen Darts

A pen shaft is suitable for dart making because the casing is fairly light and it's easy to install a pushpin in the end.

01 > Disassemble the pen. The only piece you'll need is the casing. The other components will add too much weight to the dart.

02 > Glue the pushpin inside the pen casing. If the pen casing is irregularly shaped, insert the pushpin into the heavier end.

03 > Cut 2 or 3 small triangular fins from the card stock and tape them to the back end of the dart. Tape each fin on both sides to make sure it stands perpendicular to the dart shaft. Fins are necessary to stabilize the dart. The straighter the fins, the straighter the dart will fly. Use materials that are light yet rigid. If you use card stock or cardboard that is too heavy, it will pull the dart off balance. It is best to have most of the weight toward the front.

04 > Wrap several rounds of wire around the front of the dart. This creates a leading weight, which will help keep the dart pointed forward. (See page 52 for more information about this crucial factor.) If wire is not available, fill the tip of the pen casing with hot glue.

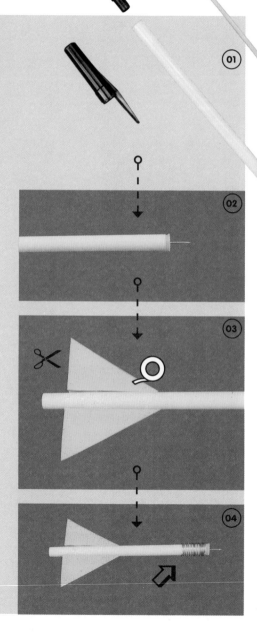

Pencil Darts

Pencils are not ideal dart material because they are dense and heavy, but they can be used in a pinch.

01 > Hot glue a pushpin to the eraser. It might seem counterintuitive to make the eraser the front of the dart, but this end is heavier than the other, which will contribute to the leading weight.

02 > Secure the pushpin further by wrapping it with masking tape.

03 > Wrap wire or another dense material around the front of the dart to enhance the leading weight.

04 > Create fins for the dart as in step 3 of **Pen Darts.** This example uses fins made from a playing card. Make sure that the fins are straight.

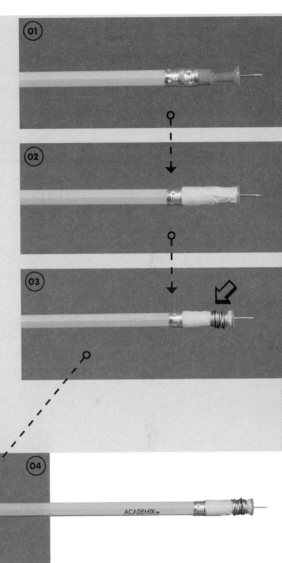

Straw Darts

Lightweight plastic straws are perfect for improvised darts because the leading weight has a profound effect on its trajectory.

01 > Glue a pushpin into one end of the straw.

02 > Further secure the pushpin with making tape.

03 > Create fins for the dart as in step 3 of **Pen Darts.** The fins shown here are made of a business card, which is very thin but also very rigid.

04 > Wrap the front end of the dart with wire to add leading weight. Secure the ends of the wire with glue.

05 > Set up a corrugated cardboard target and see what you can do. Adjust the leading weights of your darts as necessary for better control.

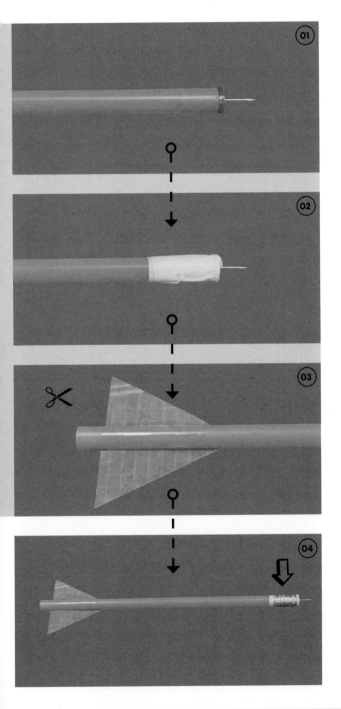

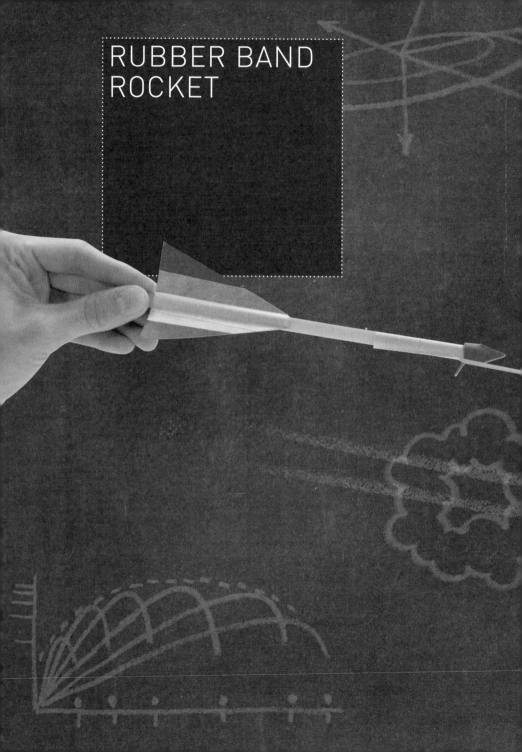

RUBBER BAND
ROCKET

ⓘ This two-part project—rocket and launcher—has a huge payout of fun for a minimal investment of time and materials. Rubber Band Rockets are simple, inexpensive, and mind-bogglingly effective. With practice, you can make them fly a distance of more than 200 feet (61 m), and though simple in design, they offer many opportunities for tinkering and improving performance by adjusting key variables. Make a few at a time; it doesn't hurt to have extras when you land one on the neighbor's roof!

MATERIAL SUBSTITUTIONS	
LARGE STRAW	**Ballpoint pen tube or 3 thin straws bundled and taped together**
PENCIL	**Craft stick, ballpoint pen, or another sturdy object at least 4" (10 cm) long**
PENCIL-CAP ERASER	**1" (2.5 cm) piece of a mini hot glue stick; 1" (2.5 cm) piece cut from the eraser end of a pencil; or other small, dense, rubbery material**
INDEX CARD	**Card stock; thin cardboard. Avoid using thin printer paper.**

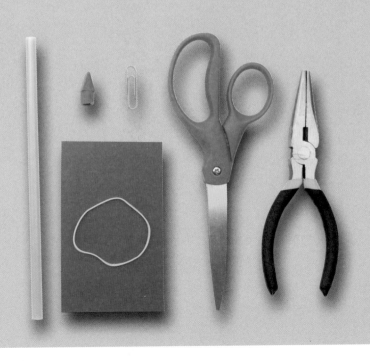

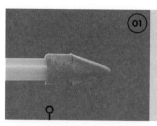

01 > Make the Rocket

Fit the eraser onto the end of the straw to weight the rocket's tip. This additional weight will help propel the rocket to greater distance and help cushion a crash landing.

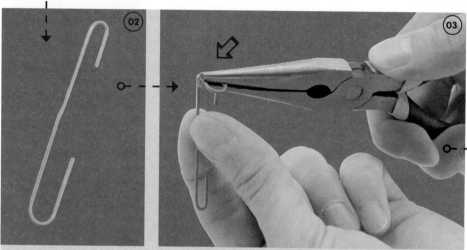

02 > Bend the paper clip into an elongated C shape.

03 > Use the pliers to bend one curved end of the paper clip into a 30-degree angle. This will be the hook for the rubber band.

04 > Position the paper clip on the straw so that the hook stands up directly below the eraser. Use masking tape to attach the remaining length of the paper clip to the straw. Wrap the paper clip in place tightly so it won't come loose when the rocket is launched!

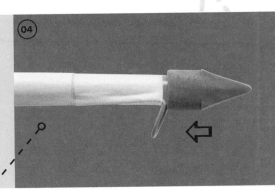

05 > Measure and cut 3 or 4 fins from the index card. Cut the fins as right angles, approximately 2" × 1½" × 1" (5 × 4 × 2.5 cm). The fins will help stabilize the rocket's flight path.

DESIGN VARIABLES: FINS

The fin size will alter the rocket's performance and maximum distance. Smaller fins will create less drag and allow the rocket to go farther, but if the fins are too small they will not stabilize the rocket's flight path. Larger fins produce more drag, but the rocket will be very stable. Very large fins create a rocket that glides gracefully. No matter the size, the fins must be aligned with the shaft to be effective.

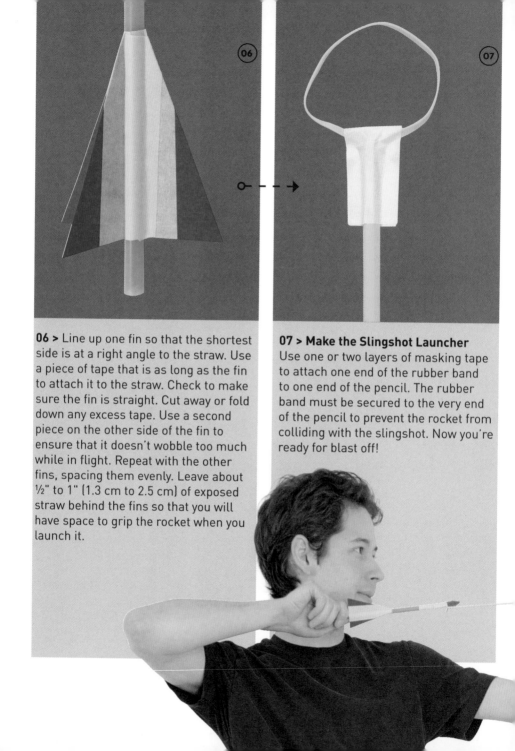

06 > Line up one fin so that the shortest side is at a right angle to the straw. Use a piece of tape that is as long as the fin to attach it to the straw. Check to make sure the fin is straight. Cut away or fold down any excess tape. Use a second piece on the other side of the fin to ensure that it doesn't wobble too much while in flight. Repeat with the other fins, spacing them evenly. Leave about ½" to 1" (1.3 cm to 2.5 cm) of exposed straw behind the fins so that you will have space to grip the rocket when you launch it.

07 > Make the Slingshot Launcher
Use one or two layers of masking tape to attach one end of the rubber band to one end of the pencil. The rubber band must be secured to the very end of the pencil to prevent the rocket from colliding with the slingshot. Now you're ready for blast off!

HOW TO LAUNCH

Hold the slingshot perpendicular to the ground with your thumb against the pencil.

Hold the rocket with your other hand, pinching the end of the rocket between the fins.

Latch the rocket's paper clip hook onto the rubber band.

Hold the slingshot out in front of you. Draw the rocket back toward yourself to build up the full potential of elastic energy. The rubber band should be perpendicular to the slingshot.

Allow the slingshot to tilt forward as you release the rocket. This will help prevent the rocket from colliding with the slingshot.

Add an extra boost to the launch by flicking the slingshot forward as the rocket is launched.

Aim higher to shoot farther, but keep the relative positions of the rocket, rubber band, and pencil the same.

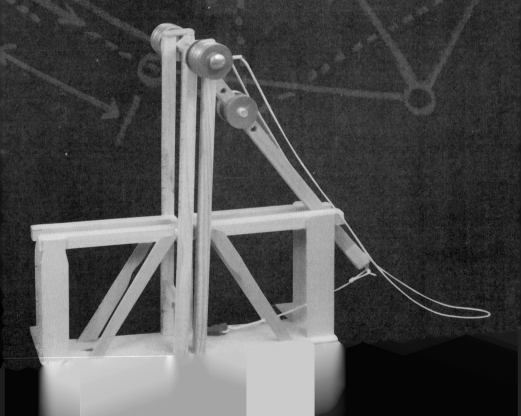

MINI SIEGE ENGINES

$$\Delta x = \frac{1}{2}at + v_0$$
$$v = at + v_0$$
$$v^2 - v^2 \, 2a\Delta$$
$$+ v_0$$
$$2$$

PYRAMID CATAPULT

ENHANCED MOUSETRAP CATAPULT

FLOATING ARM TREBUCHET

DA VINCI CATAPULT

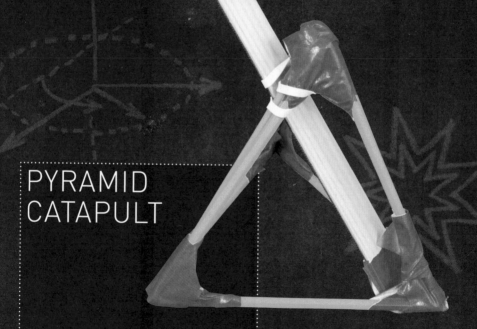

(i) Catapults don't have to be complicated to impress. This design is simple, robust, and surprisingly effective.

PYRAMID CATAPULT

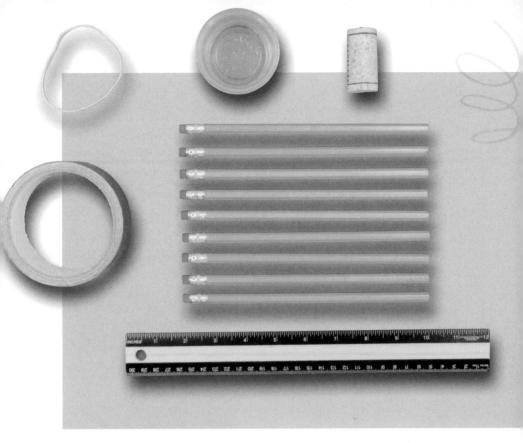

TOOLS + MATERIALS	MATERIAL SUBSTITUTIONS	
9 PENCILS OR PENS **DUCT TAPE** **THICK RUBBER BAND** **WOODEN RULER** **DISPOSABLE CUP** **CORK (PROJECTILE)**	*PENCILS*	Craft sticks, dowels, or other implements with a similar shape and strength
	DUCT TAPE	Masking tape, electrical tape
	WOODEN RULER	Paint stirrer, craft sticks glued together
	CORK	Small, dense objects

01 > Position two pencils on top of a 3" to 4" (7.5 to 10 cm) piece of duct tape. Do your best to arrange the pencils in an equilateral triangle shape. Wrap tape tightly around the ends of the pencils where they meet.

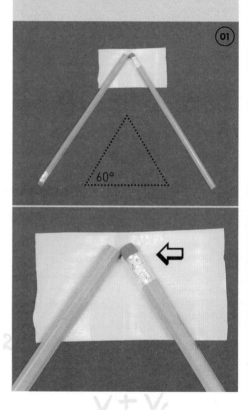

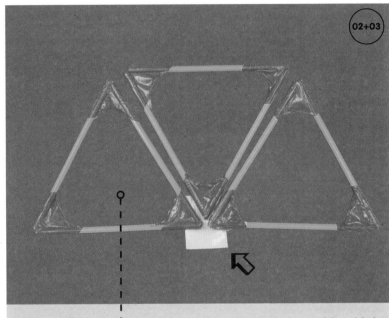

02 > Repeat this technique with a third pencil to complete the equilateral triangle. Then create two more triangles with the other pencils.

03 > Arrange the triangles so that they meet edge to edge. Use a small piece of duct tape to hold the corners of all three triangles together in the center.

04 > Position the triangles in a pyramid shape. Tape the triangles together at all four corners. You can also tape the sides of the triangles together to further strengthen their bond.

05 > Position one end of the ruler on a 3" to 4" (7.5 to 10 cm) piece of duct tape. Leave about half of the length of the tape free.

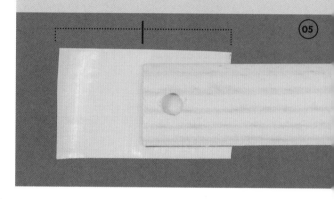

06 > Position the pyramid and ruler as shown, with one corner of the pyramid on the exposed tape. Fold the tape around both sides of the corner.

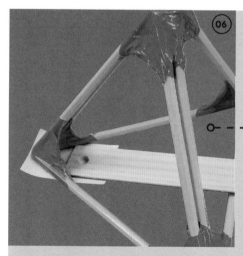

$$\Delta x = \bar{v} t$$
$$\Delta x = \frac{1}{2} a t + v_0 t$$
$$v = a t + v o$$
$$v^2 - v_0^2 = 2 a \Delta x$$

The size of this design is highly variable. You can make miniature catapults or create a giant one. As a rule of thumb, you will need stronger materials for larger designs.

What you choose to launch is up to you. Corks have a good balance of density, aerodynamics, and safety.

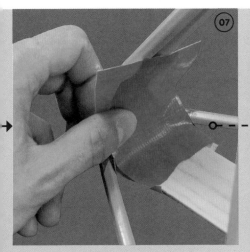

07 > Use another piece of duct tape to attach the upper side of the ruler to the pyramid. This creates a duct-tape hinge.

08 > Finish securing the hinge by folding the duct tape around the corner of the pyramid.

Use multiple rubber bands to give your pyramid catapult extra *umph*. But if you reach a point where you are unable to draw the catapult arm all the way back or your frame starts to fall apart, you've added one rubber band too many.

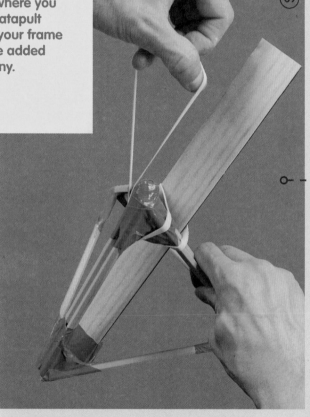

09 > Time for the rubber band. Loop the thick rubber band over the ruler, pull it through the pyramid, and then stretch it over the ruler again.

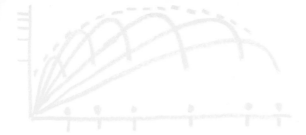

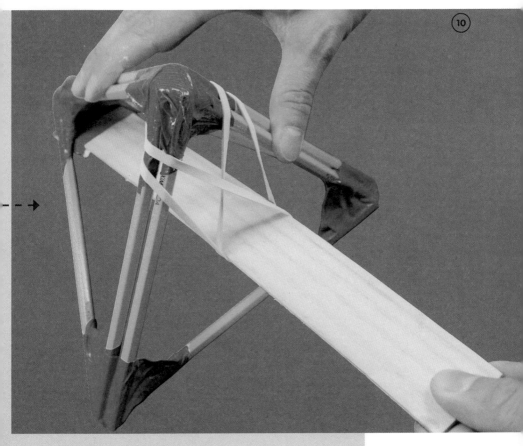

10 > The rubber band should look like this when it is attached correctly. This configuration ensures that the force of the rubber band is equally distributed.

11 > Center a 5" (12.5 cm) piece of duct tape on the back of the ruler about 1" (2.5 cm) from the end.

12 > Attach the cup by pressing the tape tightly against the sides.

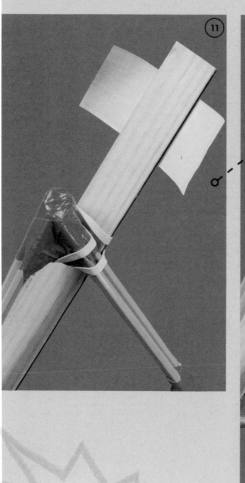

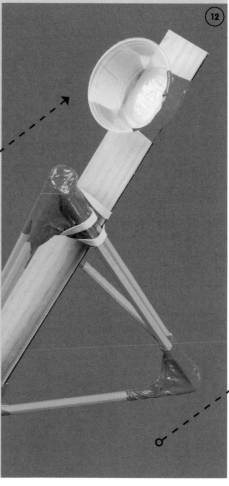

13 > To fire, take the catapult outside or into a room with high ceilings. Load your projectile into the cup. Hold down the front corner of the pyramid. Place your thumb on the exposed inch of ruler behind the cup. Pull down on the ruler as far as it will go and release it quickly.

Make sure there's nothing breakable around. You'll be surprised by how far and fast this simple design can throw!

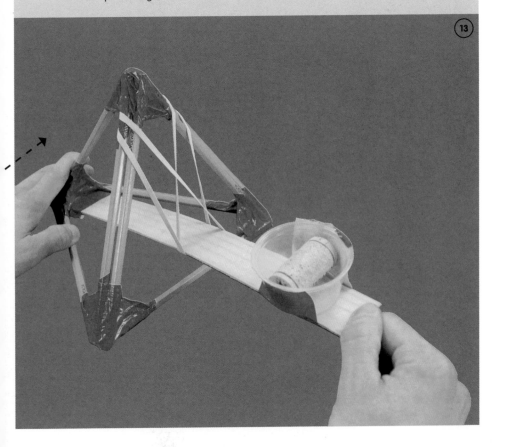

(13)

ENHANCED
MOUSETRAP
CATAPULT

$$\Delta x = \frac{1}{2}at + v_0 t$$

$$v = at + v_0$$

$$v^2 - v_0^2 = 2a\Delta x$$

$$v \frac{v + v_0}{2}$$

$$g = 10 \frac{m}{s^2}$$

(i) If you do an internet search for "mousetrap catapult," you'll find dozens of designs. Sometimes the engineering is clever, but mousetrap catapults often fail to impress for two reasons: the mousetrap's torsion spring has a limited and unchangeable amount of potential energy, and the angle of the catapult is static. The Enhanced Mousetrap Catapult utilizes the torsion spring, but also increases the amount of energy with extra rubber bands. Furthermore, this design includes a simple mechanism for adjusting the trajectory, so you can lob projectiles over high walls or fire straight into them!

**SIX 12" (30.5 CM)
PAINT STIRRERS**

MOUSETRAP

NEEDLE NOSE PLIERS

DUCT TAPE

DRILL WITH ⅛" (3 MM) BIT

RULER

MARKER OR PEN

RUBBER BANDS

UTILITY KNIFE

HOT GLUE GUN

⅛" (3 MM)-THICK BAMBOO SKEWER

SMALL CUP

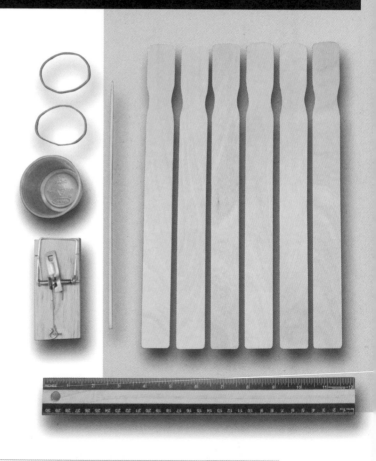

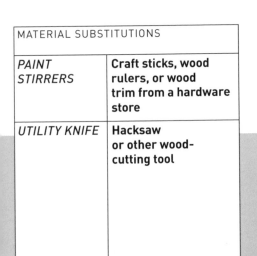

MATERIAL SUBSTITUTIONS	
PAINT STIRRERS	**Craft sticks, wood rulers, or wood trim from a hardware store**
UTILITY KNIFE	**Hacksaw or other wood-cutting tool**

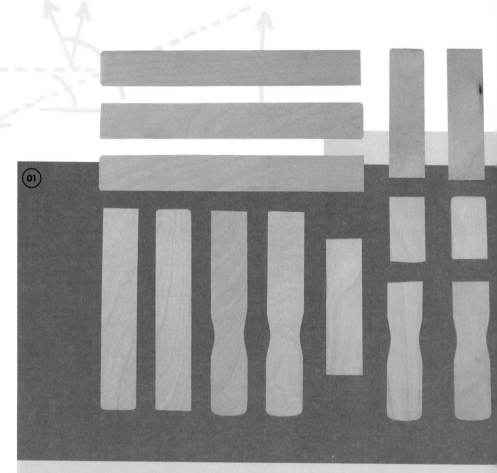

01 > Prepare the paint stirrers by cutting them into three 8" (20.5 cm), four 6" (15 cm), five 4" (10 cm), and two 2" (5 cm) pieces.

02 > Prepare the mousetrap by using the needle nose pliers to remove the trigger bar.

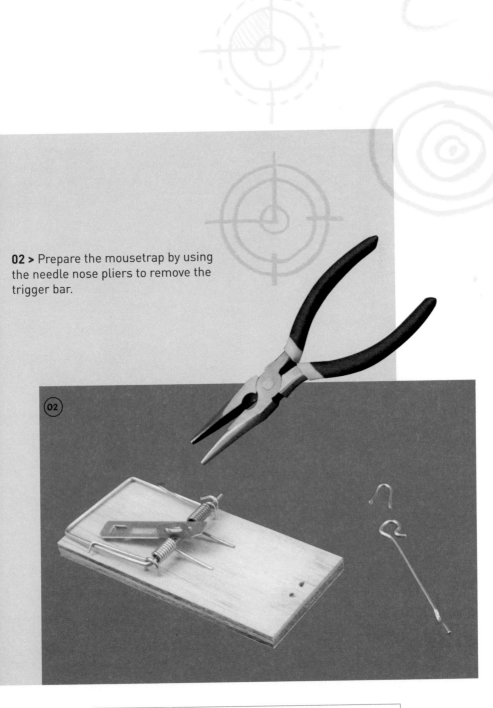

(02)

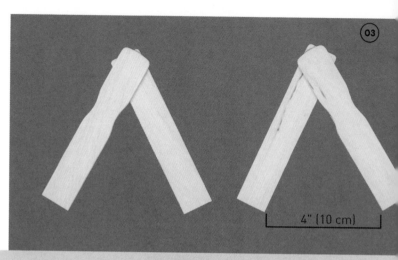

4" (10 cm)

03 > Hot glue the four 6" (15 cm) pieces of paint stirrer into two A-frame shapes. The widest part of the opening between the legs of the A-frame should measure 4" (10 cm).

04 > Stack the two A-frames and wrap the ends in duct tape to hold them together. Drill a series of holes along one leg of the A-frames, starting at the apex.

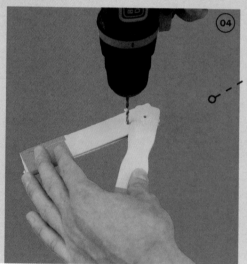

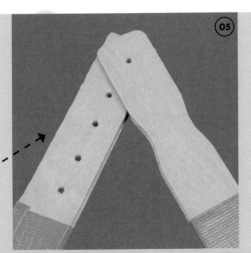

05 > Aim for five holes spaced about ½" (1.3 cm) apart.

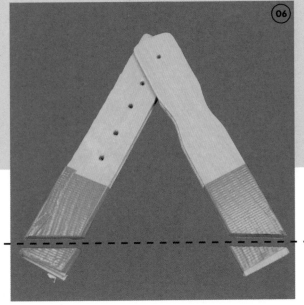

06 > Place a ruler across the bottom of the A-frames and use a marker or pen to draw a line across the legs. Trim the bottom of each leg. Using a utility knife, score the wood along the marked line several times then snap off the end with your fingers or the pliers.

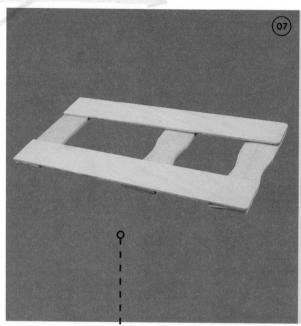

07 > Create a rectangular base for the catapult. Lay out two parallel 8" (20.5 cm) pieces of paint stirrer. Hot glue three 4" (10 cm) pieces of paint stirrer across them as shown. The two 4" (10 cm) pieces that are closest to each other should be 2" (5 cm) apart.

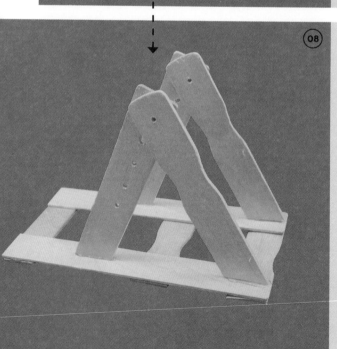

08 > Remove the tape and separate the A-frames. Glue the A-frames to the base, making sure the drilled holes are aligned. Note that the frames are positioned toward the front of the catapult where the mousetrap will be.

09 > Sandwich the mousetrap bar between the two 2" (5 cm) pieces of paint stirrer and use plenty of hot glue to attach the three layers. Allow the glue to set completely.

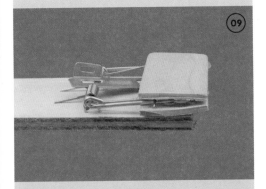

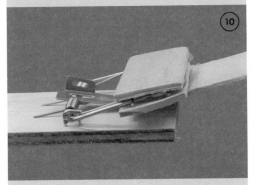

10 > Glue the remaining 8" (20.5 cm) piece of paint stirrer to the under part of the sandwich you just created, as shown.

11 > Glue the mousetrap assembly to the front of the catapult with the catapult arm tilted forward.

12 > Lift the catapult arm and insert the bamboo skewer through a pair of drilled holes in the A-frames. This will keep the arm in place while you finish the catapult.

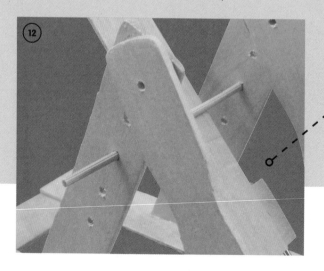

13 > Glue the two remaining 4" (10 cm) pieces of paint stirrer to the front of the A-frame for stability.

14 > Place a piece of duct tape on the back of the catapult arm, about 1" (2.5 cm) from the end.

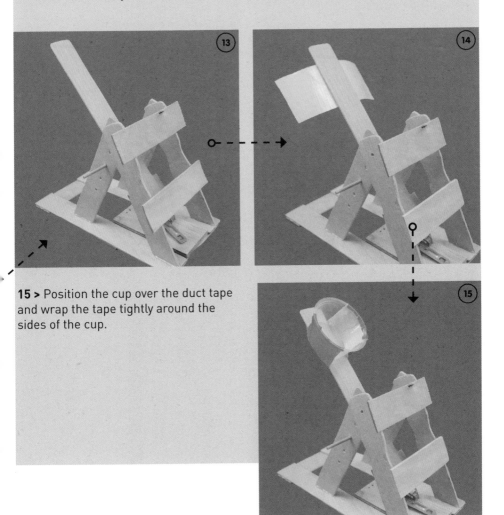

15 > Position the cup over the duct tape and wrap the tape tightly around the sides of the cup.

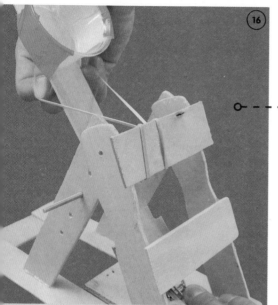 

16 > Use a hitch knot (slipping one end through the other) to attach a rubber band to the frame brace then loop the rubber band over the catapult arm.

17 > Attach a second (or a third or fourth) rubber band for added power!

18 > The trajectory of the projectile is determined by the angle of the catapult arm when it's at rest. You can adjust the catapult's trajectory by removing and reinserting the skewer at different levels. As a rule of thumb, the projectile will launch in the same direction that the cup is facing. In the first picture, the catapult has a very high trajectory, and in the second, it has a straight trajectory.

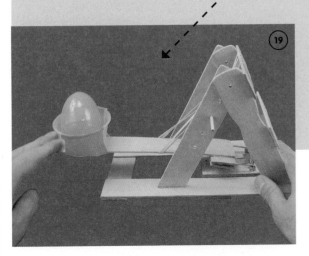

19 > Load your favorite projectile. Hold down the front of the catapult with one hand and draw back the arm with your other hand. Release by allowing your fingertips to slip off the catapult arm.

ⓘ The Floating Arm Trebuchet is a modern take on an ancient war machine. This variation features a drop channel, which allows the counterweight to fall straight down rather than swing with the arm. This is a more efficient way to transfer potential energy into kinetic energy on a small scale. A pair of wheels allows the arm to roll along glide rails as the counterweight falls. This design is more complicated to build than a traditional trebuchet, but the launching mechanism is more effective and much more gratifying to behold!

FLOATING ARM
TREBUCHET

MINI SIEGE ENGINES

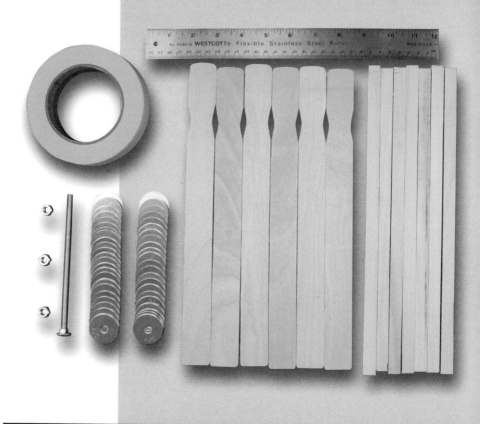

SIX 12" (30.5 CM) PAINT STIRRERS

UTILITY KNIFE

RULER

HOT GLUE GUN

DRILL WITH 1/4" AND 1/8" (6 MM AND 3 MM) BITS

SEVEN 12" (30.5 CM) SQUARE DOWELS, 1/2" (1.3 CM) WIDE*

1/4" (6 MM) BOLT, 5" (12.5 CM) LONG

SEVENTY 1/4" (6 MM) FENDER WASHERS

THREE 1/4" (6 MM) HEX NUTS

1/4" (6 MM) WOODEN DOWEL

MASKING TAPE

1/8" (3 MM)-THICK BAMBOO SKEWER AT LEAST 30" (76 CM) OF STRING

PAPER CLIP

*Square dowels must be perfectly straight or the trebuchet may not operate smoothly.

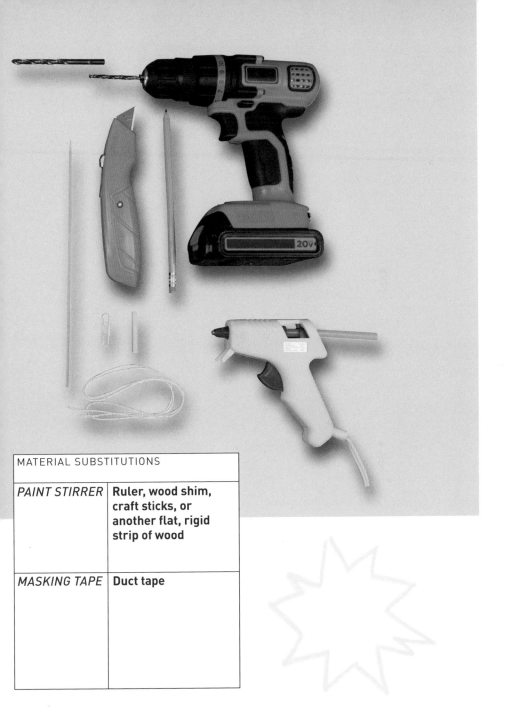

MATERIAL SUBSTITUTIONS	
PAINT STIRRER	**Ruler, wood shim, craft sticks, or another flat, rigid strip of wood**
MASKING TAPE	**Duct tape**

01 > Prepare the pieces:

Cut two 12" (30.5 cm) paint stirrers
into four 6" (15 cm) pieces.

Split two of the 6" (15 cm) pieces
down the middle.

Cut the bottom of the split pieces at
approximately a 45-degree angle.

Cut 2 paint stirrers into four 5" (12.5 cm)
pieces, leaving two 2" (5 cm) pieces.

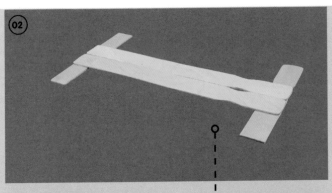

02 > Form the base using two whole paint stirrers and the two 6" (15 cm) pieces. Center and hot glue the ends of the long pieces to the shorter cross pieces.

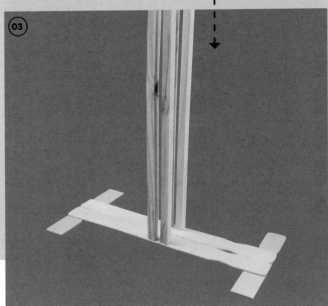

03 > Create the drop channel. Hot glue four square dowels upright at the center of the base. The gap between the pair of dowels on each side of the base is ¼" (6 mm) wide. The space between the dowel pairs across the base is 1½" (4 cm).

04 > Use the 5" (12.5 cm) pieces of paint stirrer and the four 6" (15 cm) square dowel pieces to form the glide rails.

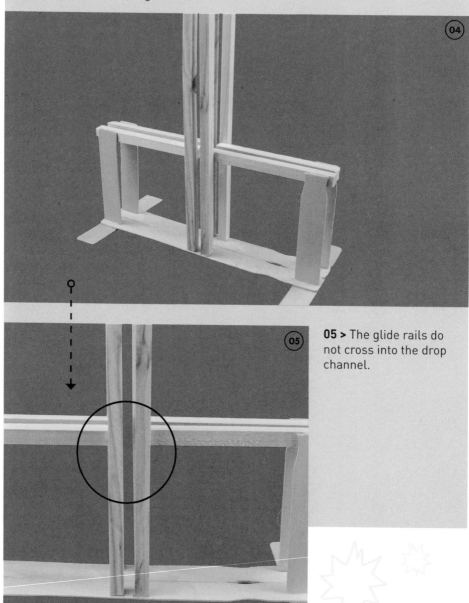

(04)

(05)

05 > The glide rails do not cross into the drop channel.

06 > Use the 2" (5 cm) pieces of paint stirrer to add structural support to the glide rails and to help maintain a uniform gap.

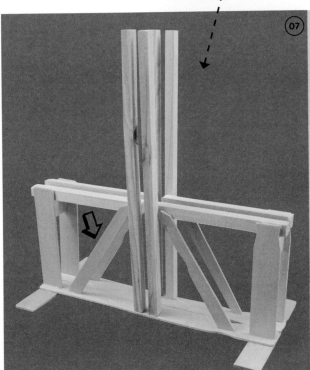

07 > Glue the split paint stirrer pieces at an angle to support the glide rails and drop channel. The tops of these supports have been trimmed for a cleaner look.

08 > Get ready to assemble the counterweight and trebuchet arm. You'll need all the hex nuts and all but 10 of the washers. The washers and nuts will be assembled onto the bolt in the order shown. Divide the remaining washers into 2 equal piles.

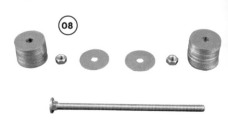

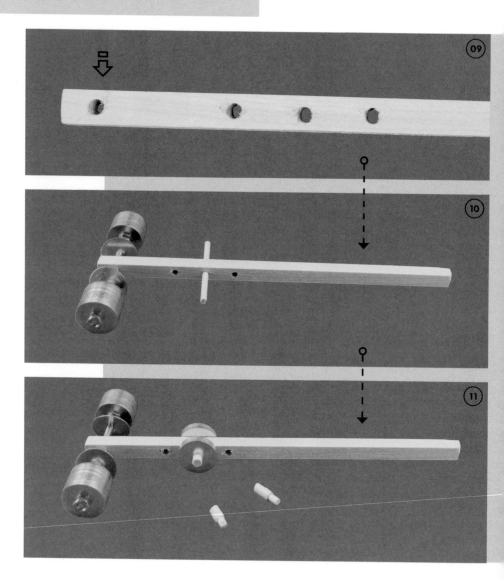

RUBBER BAND ENGINEER All-Ballistics Pocket Edition

09 > Drill four ¼" (6 mm) holes into a square dowel. One hole should be near the end of the dowel, and the other three are spaced 1" (2.5 cm) apart, starting 2" (5 cm) from the first hole.

10 > Assemble the metal components with the square dowel in the center. Insert a 2" (5 cm) piece of ¼" (6 mm) round dowel into one of the remaining holes.

11 > Wrap layers of masking tape around the round dowel until the thickness is slightly larger than the center of the washers. Snugly fit 5 washers onto each end of the dowel. The dowel should spin freely, but the washers should remain in place. Cut off any excess dowel to prevent it from getting caught on the drop channel beams. This will be the roller that travels over the glide rails.

12 > Set the counterweighted arm into the drop channel. Give it a quick test by pulling the arm back and letting it go. The arm should smoothly swing forward as the counterweight falls.

If the arm is colliding with the glide rails, carefully detach the drop channel dowels from the base and reglue them slightly farther apart.

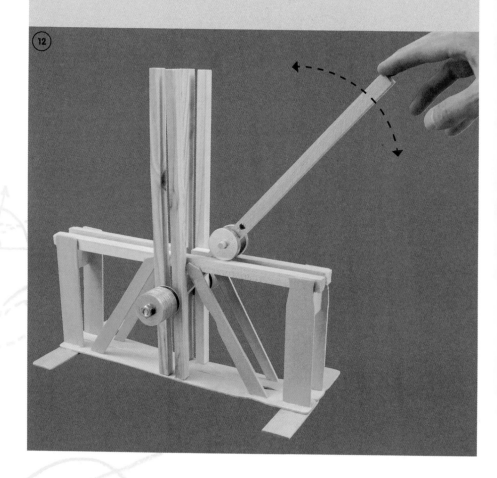

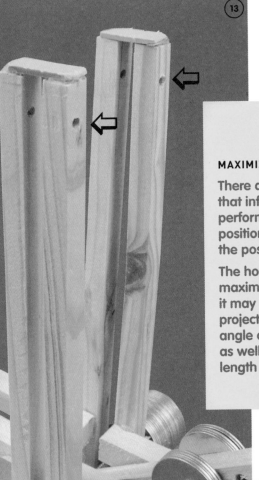

MAXIMIZING PERFORMANCE

There are several subtle but key variables that influence how well the trebuchet performs. The rollers in the arm can be positioned into different holes, thus altering the position of the fulcrum.

The hole nearest the end of the arm maximizes the speed of the projectile, but it may not perform as well with heavier projectiles. As discussed on page 108, the angle of the paper clip is a key variable, as well as the type of projectile, and the length of string to which it's attached.

13 > Once the arm and counterweight are swinging smoothly, use two scraps of wood to close off the top of the drop channel. Drill two sets of holes using the ⅛" (3 mm) bit for the trigger pins.

⑭

14 > Create the trigger by tying and taping one 30" (76 cm) piece of string to two 2" (5 cm) pieces of bamboo skewer.

15 > Set the trigger by raising the counterweight above the holes and inserting the pins. The pins should fit in the holes loosely. The goal is to use the weight of the counterweight (rather than friction) to hold the pins in place.

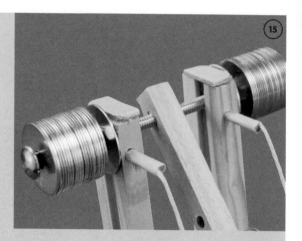

16 > Straighten out a paper clip and then fold it in half. Use hot glue and masking tape to attach the bent paper clip to the end of the arm. Bend the rounded end of the paper clip upward slightly. This is where the projectile will be attached.

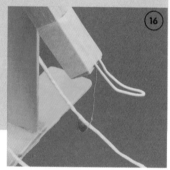

EXPERIMENT WITH THE PAPER CLIP ANGLE

The angle of the bend in the paper clip will determine the timing of the projectile's release. A straighter paper clip will result in a quicker release and a higher trajectory. A paper clip with greater curve will give a delayed release and a straighter trajectory. There's no "best" angle—it depends on the trajectory you want to achieve.

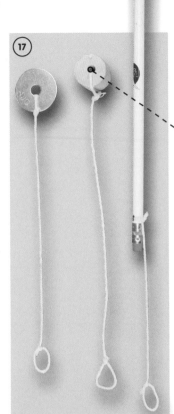

18 > Get ready to fire! Loop the string onto the paper clip and place the projectile onto the base. Stand to the side of the trebuchet to avoid being struck by a wayward projectile. Give the trigger a swift tug to release the pins.

17 > Create some projectiles from household items. All you need is a small and dense object with a string tied to it. A loop at the end of the string allows the projectile to attach to the paper clip.

DESIGN VARIABLE

The form and density of the projectile have a big impact on how well the trebuchet performs. The string length will also affect how far your projectile will launch. Experiment with a variety of items to find which one will go the farthest. Tip: Begin with small but dense items. Also, experiment with the position of the glide rail rollers to different holes in the trebuchet arm.

DA VINCI CATAPULT

$$v^2 - v_0^2 = 2a\Delta$$

$$\frac{v + v_0}{2}$$

ⓘ This type of catapult derives its name from the extraordinary inventor Leonardo da Vinci. He sketched an innovative catapult design that winds rope around a central drum and stores energy in flexing bows. Although aesthetically pleasing, many da Vinci catapult models have lacked power— until now. This take on the da Vinci catapult maximizes the amount of flexing in the bows to deliver a stellar launch while still maintaining the unique double-bow look.

MINI SIEGE ENGINES

**THIRTEEN 12" (30.5 CM)
PAINT STIRRERS**

MEASURING TAPE

UTILITY KNIFE

DRILL WITH ¼" (6 MM) BIT

¾" (2 CM) SQUARE DOWEL

HACKSAW

SANDPAPER (OPTIONAL)

HOT GLUE GUN

PENCIL

¼" (6 MM) ROUND DOWEL

STRING

DUCT TAPE

BINDER CLIP

SMALL CUP

**PROJECTILE OF
YOUR CHOICE**

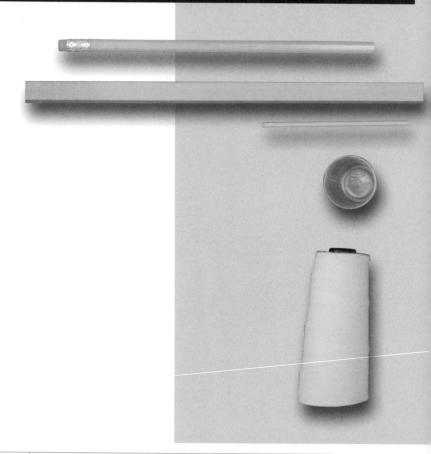

MATERIAL SUBSTITUTIONS	
PAINT STIRRER	For the bow, any semi-flexible, flat material such as a wood ruler. The other paint stirrers can be replaced with any rigid material, such as craft sticks or wood shims.
HACKSAW	Any tool that can cut the ¾" (2 cm) dowel

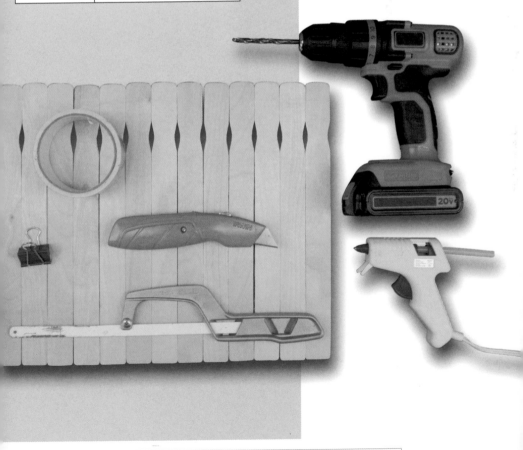

01 > Select two paint stirrers with very straight wood grain. Cut the two stirrers to 7½" (19 cm) and save the remaining pieces. With the utility knife, score the 7½" (19 cm) pieces in half lengthwise and then break apart. If the grain of the wood is straight, they should split evenly.

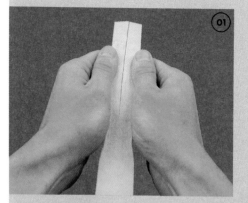

02 > Prepare the remaining pieces of paint stirrers. Cut nine 3½" (9 cm) pieces. Cut three 6" (15 cm) pieces. Drill a ¼" (6 mm) hole near the end of two 6" (15 cm) pieces.

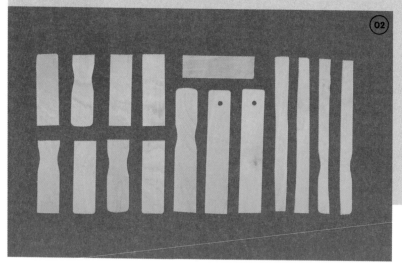

03 > Form a rectangular base, using two 3½" (9 cm) pieces of paint stirrer and two whole stirrers. Hot glue the pieces together at the corners.

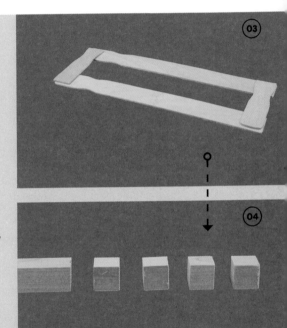

04 > Cut four ¾" (2 cm) cubes with the hacksaw. **Optional:** Clean up the edges of the cubes with sandpaper or by scraping them against another abrasive surface, such as concrete.

05 > Glue the four cubes to the corners of the base. Two of the cubes diagonally across from one another should be indented slightly from the edge to allow for the attachment of the bows in step 6.

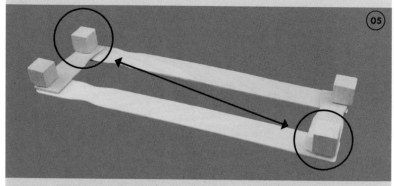

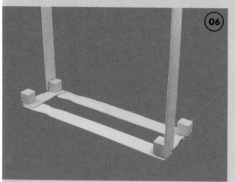

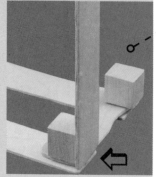

06 > Glue a whole paint stirrer vertically to the two indented cubes. The flat edge of the bow should be aligned with the edge of the adjacent cube.

07 > Position a 3½" (9 cm) piece of paint stirrer horizontally from the bottom of each bow to the next cube. Hot glue the pieces into place at each end. Line up a second bow with each of the first and hot glue them into place at the base only.

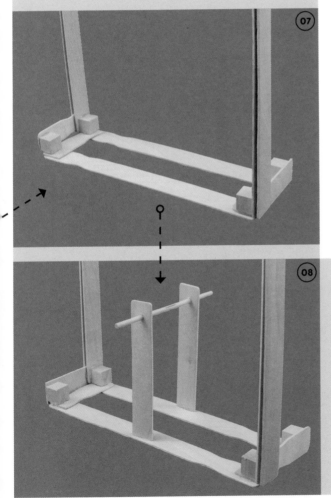

08 > Mark the center point on the long sides of the base and hot glue the two drilled 6" (15 cm) pieces of paint stirrer into place, directly across from one another. These will form the fulcrum. Cut a 5" (12.5 cm) piece of dowel. Insert the dowel through the drilled holes to make sure they're aligned.

09 > Hot glue the split pieces of paint stirrer into place as trusses to strengthen the fulcrum supports.

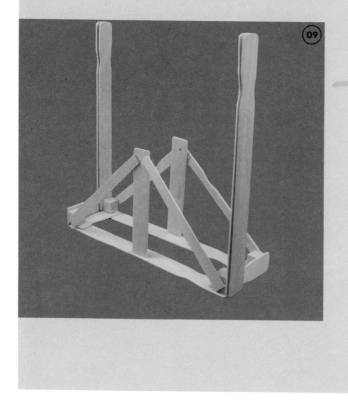

10 > Line up three of the 3½" (9 cm) pieces of paint stirrer. Hot glue two more 3½" (9 cm) pieces on top to form a pad.

11 > Hot glue a whole paint stirrer to the center of the pad.

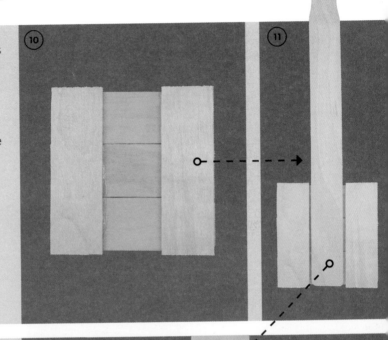

12 > Turn the pad over and use plenty of glue to attach the fulcrum dowel. The dowel should be centered on the pad.

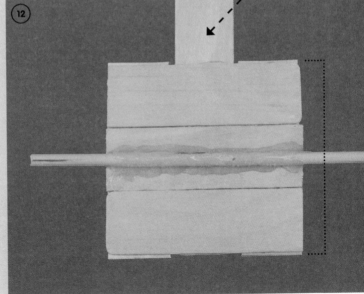

13 > Turn the pad dowel-side down. Carefully pry the fulcrum supports aside and insert the ends of the dowel through the drilled holes. The pad should fit snugly between the fulcrum supports, but there should not be any excessive friction that might hinder the catapult's performance.

14 > Cut a 45" (1 m 14.5 cm) piece of string. Fold it in half and tie a hitch knot around one side of the pad. Make sure that the string is hitched at the edge of the pad nearest the catapult arm. Attach the string to the pad with hot glue.

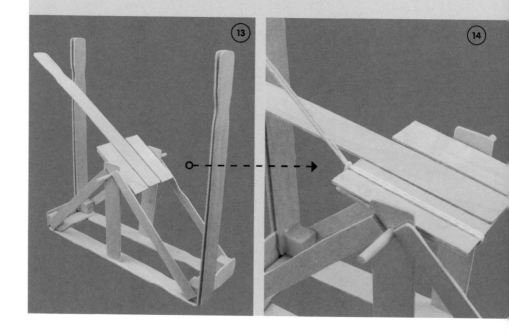

15 > Slightly bend the bow that corresponds to the string placement and fold the string over the top.

16 > Cover the string with hot glue while you hold it in place. When the glue has dried, wrap the glued string in place with duct tape to further secure it.

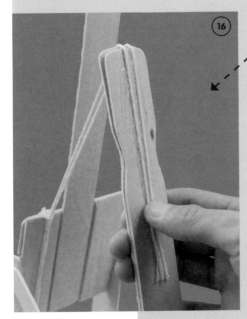

17 > Repeat steps 14 through 16 with another piece of string on the other side of the pad. This time, make sure the string is hitched at the bottom edge of the pad.

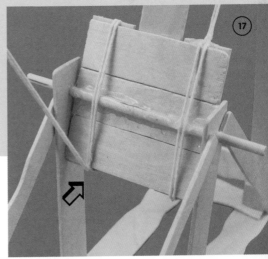

18 > Install the catapult's stop bar by bending the arm back and using the binder clip to attach the remaining 6" (15 cm) piece of paint stirrer.

19 > Use glue and tape to attach a cup to the end of the catapult arm. Leave about ½" (1.3 cm) of paint stirrer exposed at the end.

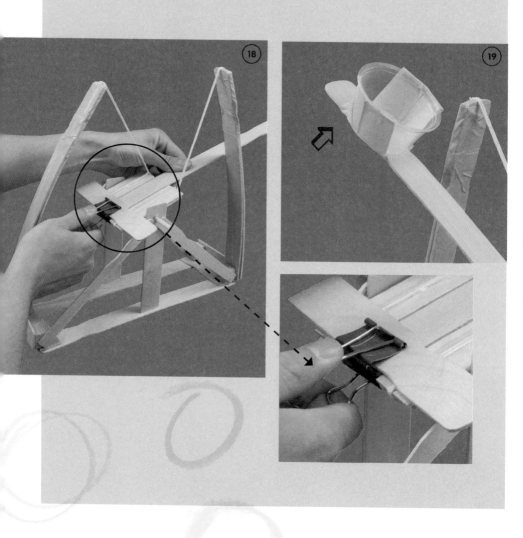

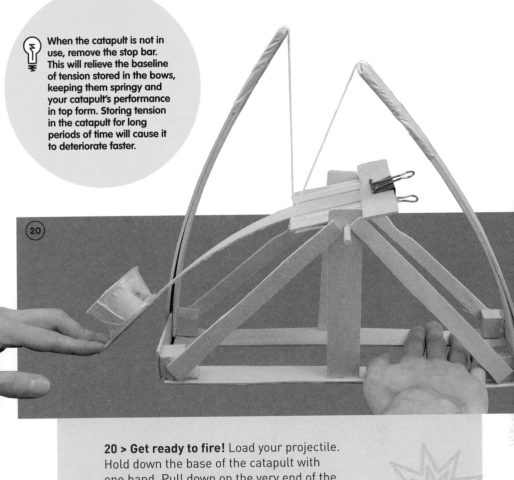

When the catapult is not in use, remove the stop bar. This will relieve the baseline of tension stored in the bows, keeping them springy and your catapult's performance in top form. Storing tension in the catapult for long periods of time will cause it to deteriorate faster.

20

20 > Get ready to fire! Load your projectile. Hold down the base of the catapult with one hand. Pull down on the very end of the catapult arm with your other hand and—**release!**

MATERIAL SOURCES

The materials used in this book are inexpensive and easy to come by. Materials not listed here can most likely be found at hardware and department stores or in your kitchen junk drawer. Note: Online vendors may not ship to all countries.

CRAFT CUBES
Cubes with holes and dowels of many varieties can be purchased at www.CraftParts.com.

CRAFT STICKS
Find crafting sticks at craft or dollar stores.

HOT GLUE
Available at crafting, hardware, or department stores. I recommend buying the glue sticks online or at dollar stores to save money.

LONG RUBBER BANDS
These are available by the box through office supply stores.

PAINT STIRRERS
Paint stirrers are available at hardware stores, sometimes for free if you ask nicely. You can also purchase them online. I bought boxes of 100 for this book for about $20 USD.

ACKNOWLEDGMENTS

A heartfelt thanks to my parents, Nancy and James, for supplying my childhood with a bounty of things to build with and old VCRs to take apart. Another big thanks to Ali, who supports me endlessly and doesn't mind at all when I take over our dining table with glue guns and craft supplies.

ABOUT THE AUTHOR

Lance Akiyama combines tinkering and education into one aspiration: to create a better world by inspiring the next generation of innovators with exciting hands-on projects.

To that end, he has created many project-based learning tutorials on Instructables.com, started a small, after-school engineering service, and is currently employed as a STEM-based curriculum developer for Galileo Learning.

Other than that, Lance spends his free time designing elaborate plans for advanced contraptions, keeping journals in cryptic backwards writing, and attempting to fly by strapping paper wings to his arms and leaping from rooftops.

INDEX